THE HISTORY OF THE

# INDY 500

# THE HISTORY OF THE

# INDY 500

**BILL HOLDER**
Photography by BOB FAIRMAN
**Foreword by JOHNNY RUTHERFORD**
Three-time winner of the INDY 500

**Brompton**

First published in 1992 by
Brompton Books Corp
15 Sherwood Place
Greenwich, CT 06830



ISBN 0-86124-741-8

Printed in Belgium

**Page 1:** *Rick Mears is one of the real heroes of the Indy 500, having won it four times. This is his 1984 win.*

**Previous pages:** *Tero Palmroth (56), Didier Theys (12) and Emerson Fittipaldi (20) duel for position during the early stages of the 1989 Indy 500. The race was won by Fittipaldi with Al Unser, Jr. finishing second after a late-race crash.*

**Below:** *A special superspeed filter captures the essence of high speeds at Indy. This is Emerson Fittipaldi during the 1985 race.*

**Below right:** *Danny Sullivan won the 1985 race under strange circumstances. While duelling with Mario Andretti, Danny looped the car 360 degrees, but was able to recover and win his first and only Indy 500.*

# CONTENTS

**ACKNOWLEDGMENTS**

The author would like to thank the following people for their assistance in the preparation of this book: Deke Houlgate of Pennzoil Public Relations; Jean Martin, the editor; Rita Longabucco, the photo editor; Don Longabucco, the designer; and Elizabeth A. McCarthy, the indexer.

**PHOTO CREDITS**

Photography by Bob Fairman except for the following:
Bob Fairman Collection: 10(top), 13(both), 38(top).
Bill Holder: 49(top), 52-53, 57.
Bill Holder Collection: 12.
Lou Lengerich: 20-21, 61, 70.
John Mahoney: 37(top), 38(bottom), 44(bottom).
Courtesy of Pennzoil: 7, 34(bottom), 39(top), 47(bottom left), 67, 71(bottom).
UPI/Bettmann: 10(bottom), 11(both), 14, 17, 36, 44(top), 72-73.
Bill Zmirski: 2-3, 8(right), 9, 15, 16-17, 33(bottom), 35, 37(bottom), 40(bottom), 41(bottom), 47(top), 54, 55, 56-57, 66(bottom), 68-69, 76, 77.

# FOREWORD

I've never stood waiting to receive the kickoff to start the Super Bowl, but how could that compare with accelerating toward the green flag that starts the Indianapolis 500? I've started the race up front, and I've started it near the back, and I can't imagine any feeling like the rush to the start of the Indianapolis 500 in one of the world's great racing cars.

Once the race is under way, I'm too busy to have any feelings at all. But the buildup to the Indianapolis 500 is an emotional experience that has no equivalent in any other sport.

The ultimate emotional high comes when you win the race, and to tell the truth even with the experience of winning again and again, every time on the victory stand feels just as exhilarating as the first. I recall the first, second and third wins now with the same sensations of thrill, personal satisfaction and gratitude for the sacrifices of my team as I did on the podium years ago.

The legacy of the Indy 500 is that it's a one-of-a-kind event, impossible to duplicate anywhere else. It's more than a sporting event, more than an auto race. Winning at Indianapolis is the ultimate victory, enjoyed with just as much relish by the mechanics, the car owners, the wives, the friends, the family, and the loyal fans as it is by the driver.

Johnny Rutherford
Indianapolis 500 Winner
1974-1976-1980

# BEGINNINGS

Indianapolis, Indiana. No other city in America, or in the world for that matter, has such an association with a sporting event as does Indy. The association, of course, is with the greatest spectacle in racing, the Indy 500.

Held on Memorial Day weekend, on the last Sunday in May, the Indy spectacle is the largest single sporting event of any kind in the world. It attracts over 400,000 fans every year, and is broadcast to every corner of the globe. More than just an auto race, it is a happening of monumental proportions. The world's business community gathers in elaborate hospitality centers on the famous old track's grounds to close business deals and to socialize with the rich and famous. Movie stars and government heads wander around the huge grounds, totally enthralled with the entire scene, just like other fans.

Race day is, of course, the big day, but it concludes almost a month of speed activities building up to the big event. There are four days alone of qualifications on the two weekends preceding the race itself. Three days before the race is the historic Carburetion Day (even though the cars no longer use carburetors) when the whole starting field is given two hours for final tune-up and race day set-up work.

The significance of the event cannot be overestimated. The Indy 500 is the only example in racing where the name of the event is used to identify the type of cars that run there. The term "Indy Car" identifies the low-slung screamers that run at the famous two and-a-half mile oval, even though the cars run at many other tracks throughout the season. The Indy's significance is also in evidence in other types of racing. The Daytona 500 for NASCAR stock cars, for example, is often described as "The Indy 500 of Stock Car Racing." That's just about the ultimate honor that can be paid to a racing event.

Much of the significance of the great race (that's what the Indy 500 is often called and nobody ever asks which race is being talked about) has to do with its longevity. The race in 1991 was the 75th running of the event.

Every year the speed increases – and every year observers pronounce that the race has reached the ultimate speed that can be run on the mildly-banked track. In fact, those words were first heard after the running of the initial event way back in 1911. The local papers indicated that, at 74 miles per hour, the speed limit had been reached for the grueling affair. To run this track any faster would be just too much for man and machine to bear. Today, speeds reach in excess of 220 miles per hour!

With the exception of six years (1917-18 and 1942-45) when the world was at war, the event has been run every year. A little known fact, though, is that there was

**Below left:** *Didier Theys prepares for the 1990 race.*

**Right:** *Moving through turn four, Emerson Fittipaldi acknowledges the cheering crowd after his 1989 victory. Fittipaldi is a former World Champion in Formula 1 road racing.*

**Below:** *The front row in 1988: Rick Mears (5), Danny Sullivan (9) and Al Unser, Sr. (1).*

20
Marlboro

**Below left:** *The first driver to win the Indy 500, Ray Harroun covered the distance in six hours 42 minutes and eight seconds at an average speed of 74.59 miles per hour. For the time, it was an amazing feat.*

**Right:** *Push-off for the first pace lap occurs at the 1926 Indy 500. The race was won by number 15, driven by Frank Lockhart at an average speed of 95.904 miles per hour.*

actually a race on the track in 1909. Hardly 500 miles, that first race long ago was only five miles in length, and resulted in five deaths.

But let's step back in time to that first 500-mile race in 1911. Wouldn't you have just loved to have been there? The crowd was clad in derby hats, and spindly bleachers held fans who, though fewer in number, were probably just as avid as those in the 1990s. The air was filled with acrid smoky exhaust. And each car, except winner Ray Harroun, carried both a driver and a riding mechanic.

That famous winning number 32 Marmon Wasp has been captured through the years in porcelain and paint. An interesting innovation that winner Harroun incorporated was the mounting of a primitive "rear-view" mirror, as there was no mechanic to turn around and tell him the situation behind him. Through the years, Indy has been responsible for a number of innovations that have made their way into passenger cars.

It took real men to drive those rough high-riding machines for what seemed like unending hours back in those days. The tenth place car in the 1912 race, for example, took almost nine hours to cover the total distance. The driver even had lunch along the way!

Those early cars used huge engines of nearly 500 cubic inches. But several years into the race, those cubes started dropping to more reasonable values. During the late 1920s, the values were as low as only 90 cubic inches. The trend then moved back up to the 200 and 300 cubic inch level in the 1950s and 1960s, and finally stabilized at 161 cubic inches in 1977, where it has stayed ever since. Also during those first years, European companies including Mercedes and Peugeot dominated the scene.

During the 1920s, the track was taken over by the colorful war hero Eddie Rickenbacker, and Indy flourished. The decade also saw a burst of speed-producing technology in the form of new front wheel drive systems and the addition of superchargers for increased horsepower. The improvements resulted in the race time falling below the five hour mark. In 1925 the race (won by Peter DePaolo) averaged over 100 miles per hour for the first time.

The Great Depression during the early 1930s caused the engine displacements to be raised in order to reduce

**Left:** *The starting field for the inaugural Indy 500 in 1911 fires their monster engines before the start of the race. The engines were huge by today's standards, displacing nearly 600 cubic inches. The smoke from the motors and dust from the track made visibility for all but the front row very difficult.*

**Right:** *Wilbur Shaw wins his first of three Indy 500s in 1937, beating Ralph Hepburn by less than 20 car lengths. He set a new speed record for the race of 113.580 miles per hour, more than four miles per hour faster than the previous standard established the year before by Lou Meyer.*

15
15
10
39

costs by allowing stock car engines to be modified and used. Indy winners were now becoming household names and international celebrities, with such drivers as 1934 winner "Wild Bill" Cummings and three-time winner Wilbur Shaw enjoying that status.

With the world scene churning in the early 1940s, 1941 would be the final shot at the famous track for five long years. The track sat unattended as Mother Nature worked on taking it back. In fact, there was a time following the war when it appeared that there might not be another race due to disinterest in refurbishing the overgrown track. Had Indianapolis businessman Tony Hulman not purchased and restored the track, the great race could have died right then.

The 1950s could well be considered the start of the modern era at Indianapolis. It was a time when cars started looking similar with what was called an upright design. It was a big change from earlier years when there were all kinds of body styles and engine configurations. The cars of the 1950s sported long sleek noses with the engines mounted beneath.

Famous names abounded during the decade when rock-and-roll was born. Bill Vukovich won two races, but was killed trying for number three. Twenty-two-year-old Troy Ruttman became the youngest driver ever to win in 1952.

During the 1960s, one of the greatest revolutions ever to hit the old Hoosier State track came over from

**Left:** *The 1953 Ford convertible pace car leads the field through turn one on the parade lap. On the pole is number 14, Bill Vukovich, who would win the race in 1953 and 1954 but lose his life in a crash while leading the 1955 Memorial Day classic.*

**Below:** *The release of thousands of colorful balloons signals that the start of the race is moments away.*

**Bottom:** *Briton Jack Brabham revolutionized Indy Car design by introducing this lightweight rear engine concept in 1961.*

**Left:** *Al Unser, Sr. celebrates his third Indy 500 win in Victory Lane in 1978. He would win again in 1987 to become the second of three four-time Indy winners.*

**Above right:** *Ludwig Heimrath, Jr. (71) and Pancho Carter (24), who finished 13th and 22nd respectively, play follow-the-leader during the 1989 race. Heimrath got his racing start in go-karts in Canada in 1972 and was the 1976-77 national kart champ. Pancho is the son of former Indy 500 driver Duane Carter. Carter started in quarter midgets as a youngster and has won USAC titles in midgets, sprint cars, and dirt cars.*

Europe with the introduction of the rear engine race car design. Englishman Jimmy Clark proved that having the engine located in the rear was the way to go when he won the race in 1965. It set the trend for the years to come, as the Indy Car sport quickly turned that way following Clark's success.

The decade also saw the debut of one of the most famous drivers ever to drive on the famous track: A.J. Foyt won the race twice during the 1960s, in 1961 and 1967. Other famous winners of the decade included Mario Andretti, Jim Rathmann, and Parnelli Jones.

Indy is a dangerous place, and many men have died over the years. One of the worst incidents took place in 1964, when rookie Dave McDonald and Eddie Sachs were killed in a crash on the main stretch.

The 1970s saw the beginnings of the technical revolution at Indy. It came in the form of huge wings, both front and rear, and giant power from the engines. The decade started with two straight victories from Al Unser, Sr., who would take his third checkered flag during the same ten years with his third victory in 1978. A.J. Foyt won what will probably be his last 500, his fourth, taking the victory lap in 1977. Johnny Rutherford won two of his three Indy victories during the 1970s, taking the 1974 and 1976 races.

The 1970s was a decade of initial victories that would be added to later, or victories that were added to already existing wins. Gordon Johncock won his first of two wins in 1973, Bobby Unser took his second of three Brickyard triumphs in 1975, and Rick Mears, in 1979, took his first victory of four. The 1970s also saw the first 200 mile per hour lap at the ancient facility.

The speeds continued to climb during the 1980s, and attempts were made to slow down the cars. Johnny Rutherford would win his third, and final, Indy in 1980. Wins number two and three would come to Rick Mears in 1984 and 1988, and Al Unser, Sr. would get number four in 1987. Several first time winners would become household names, like Bobby Rahal, Danny Sullivan, Tom Sneva and Emerson Fittipaldi.

The qualifying speeds continued to climb during this decade, over the 220 mile per hour mark, and there didn't seem to be a limit on how fast these cars could run around the old track. New technology kept the cars going ever faster, even though continuing attempts were made to slow them.

With the advent of the 1990s, it was more of the same with an old familiar name in Rick Mears winning his fourth in 1991. A newcomer, Dutchman Arie Luyendyk, surprised the world by taking the win in 1990.

What the remainder of the 1990s will hold at Indy is anybody's guess. But if history is any indication of things to come, the excitement, tradition, and technology will continue at the Indianapolis 500.

# THE TRACK

The shape of the old track hasn't changed since its construction some eight decades ago. The track is narrow when compared to more recently constructed superspeedways, its slightly banked turns contrasting with the high-banked high-speed tracks of modern times.

Although most closed tracks are referred to as ovals, the track at Indy is more of a rectangle, with four almost identical quarter-mile turns. They are connected by one-eighth-mile short straightaways and five-eighth-mile front and back straightaways. The track has presented the ultimate challenge to drivers and teams since it was first laid out.

The Indianapolis Motor Speedway wasn't initially conceived as a racing facility. Designer Carl Fisher thought that besides racing, the giant facility he was proposing would be of great benefit to the fledgling car industry. Along with three other local businessmen, he formed the Indianapolis Motor Speedway Corporation in 1909.

A site encompassing a massive 320 acres northwest of downtown Indianapolis was chosen. The surface selected was far from the glass-smooth surface that now coats that historic two-and-a-half miles. A combination of asphalt and crushed rock was supposed to provide a solid base, but in the inaugural race in 1909, the track broke up badly.

A change had to be made if racing was to continue. So, as was the custom for city streets at the time, the track was covered by over three million bricks. Having one of those bricks today as a collector's item is a prize for any Indy fan.

As racing at ever-increasing speeds became the standard, the need for smoothness on the track soon became evident, and in 1937 the turns were surfaced with asphalt. A year later the black surface continued to cover the bricks with only the middle portions of the main stretches remaining brick. Eventually, the whole track would be blacktopped with the exception of a "yard of bricks" at the starting line. It will always remain there as a fond memory of those exciting earlier times.

One of the signatures of the original speedway were the oriental-style pagodas, both of which stood along the main stretch. The first structure was built in 1913 and was destroyed by fire some dozen years later. In 1926, the second pagoda was constructed on the same location. It lasted until 1957, when it was torn down and the present Tower Terrace control center was built.

The biggest changes to the Indy scene have been outside the track itself, with the continuing progression of the building of bleachers, and the building of more bleachers. Along the main straightaway, additional stands started appearing in recent years both inside and outside the track. They continued through the years to

**Left:** *The flatness of the track can be seen in this shot of the 1983 front row behind the famous "yard of bricks."*

**Above:** *The cars in the 1983 race roar into the slightly banked 13-degree first turn, as the huge crowd looks on.*

run the length of the main stretch in double-decked configurations.

To witness the greatest spectacle in racing from one of the best seats is not inexpensive. The most expensive seats for the 1992 race, in the so-called Penthouse, are priced at a cool $100 a ticket. There are also "cheaper seats" priced at $45 and $55. The "general admission" seats are in the $25 and $35 range.

The rock-bottom price to get into the race, a paltry $15, is for those who are looking to have a good time in the infield, many of whom are totally oblivious to the action on the track. That's a pretty small price to pay to be this close to the action and also be a part of this huge happening.

Even in the 1990s, the construction still continues all around the track. Perhaps the most noticeable additions are the luxurious hospitality suites, used by corporations to wine and dine customers. For the 1991 race, there were 38 new suites added above the inside stands along the main stretch. It's one of the few places at the track on race day where one might see people wearing their Sun-

**Left:** *The modern control tower replaced the original oriental-style pagodas in 1957. The modern scoring pylon at the starting line provides lap and position data to the fans during the race.*

**Below:** *The thrill of an Indy victory: 1985 winner Danny Sullivan makes his way through a throng of admiring fans to take a final victory lap around the famous two-and-one-half mile oval.*

**Right:** *Luxury penthouse boxes overlook the pits and the main straightaway.*

**Below right:** *The main entrance to Gasoline Alley is where the cars are pushed to and from the pits. This arrangement allows thousands of fans to get a close-up view of the cars and the drivers.*

day best while the majority of the crowd is dressed for fun and frolicking.

The managers of the track are also always thinking about the convenience and comfort of all the fans that come to the race. A new handicapped area, capable of accomodating some hundred fans in wheelchairs, was recently completed at the south end of the track.

But probably the most recognizable structures at Indy have always been in the area known as Gasoline Alley. This is the garage area which houses the cars throughout those wonderful 30 Days in May, as the newspapers like to call the whole Indy undertaking.

The complex of 88 garages (the number actually runs to 89, with the number 13 omitted) is nestled just inside the middle of the main stretch. The complex is just a short push from the pits. The location presents a congested situation in which the teams must get their race cars to and from the track.

Over the years, there have been thoughts about building a pedestrian bridge over the area, but they have always been discounted. Old-timers around the area say that long-time owner Tony Hulman liked the situation, since it allowed the fans to get up close and personal with the cars and the drivers as they rolled by. Even today, that congestion is accepted by all as a part of the charisma of the old Brickyard.

The first of the garages were built in 1915. Prior to that time, the cars were worked on in garages in downtown

**Left:** *A Tom Sneva crewman works on a car in front of one of the old garages in 1984. These vintage garages were replaced with the new modern garages in the late 1980s.*

Indianapolis. Six months before Pearl Harbor, a mechanic's carelessness caused a huge fire in the area which destroyed a number of the garages along with two race cars.

Hulman then purchased the sadly-deteriorated track following the war for about $750,000, a definite bargain even then. The burned-out garages were replaced and, through 1967, a number of new garages were constructed. An interesting footnote is that during the early years of the race, the garages housing the cars of foreign race teams were located in a different part of the track.

Many teams retained the same garage or group of garages for many years. Records vanish and memories fade on such things as garage assignments, but many felt that long-time Indy Car builder A.J. Watson might have held the record in Garage 19. A.J. Foyt was in Garage 29 since he started his own operation. The Penske team had well over a decade in garage numbers 75-77.

Former riding mechanic Clay Ballenger said that at one time there was an opening between the back-to-back garages. "You could climb up and look over into another garage and see what they were doing," he explained.

Through the years, some teams put extra time and effort into personalizing their tiny stalls. By 1985, the Penske and Beatrice garages were spruced up beyond belief. The three Penske cars of Al Unser, Sr., Danny Sullivan and Rick Mears sat on a checkerboard floor. Mario Andretti's Beatrice Foods garage employed the services of an interior decorator to create its flashy red-and-white appearance.

In the late 1980s, the old garages finally came down, much to the sadness of many Indy fans. The old wooden structures were replaced by modern facilities which have more of a look of an industrial park than of a race car garage area.

Many fans that come to the track probably never realize that there is a golf course located along the backstretch. It's used just like any other course, but several times during May it becomes a parking lot.

The huge facility at the corner of 16th Street and Georgetown Road is as much a part of the Indy mystique as the race itself. And although the track's appearance has changed through the years, it will always be the site of the greatest spectacle in racing.

**Below:** *Lack of room in the old garages often forced crews to accomplish the major teardown and rebuild operations outside, as seen here.*

# THE CARS

The stars of the Indy 500 have always been the cars, and they have evolved through the years much like the passenger car and aircraft. Just when it seemed that they couldn't get any sleeker or more powerful, those steady steps of technology kept moving along providing the answers for more and more speed.

The evolution of the Indy Car through the years at the Brickyard can be compared with the evolution of fighter aircraft in this country. This comparison can be drawn from over a half-century ago, when the huge radial engines of the 1930s fighters dictated the need for large bloated-looking fuselages. That same trend was the rule with Indy Cars of the time, the bodies being long and wide and the drivers sitting high. It wasn't until about 1940 that the cars began to assume a more rounded and streamlined look.

The 1950s Indy Cars continued that aerodynamic trend, with sloping noses and the cars squatting closer to the track. During this step in the development time period, the cars became coined with the name of "roadster." The low-slung machines had greater speed on the turns with the engines mounted on the left side of the car, and the driveshaft whistling past the driver's right side.

A laid-over engine in the 1957 Belond Special provided the desired super-low profile and a low center of gravity. The great reduction in aerodynamic drag of that two-time winner at Indy was a signal of things to come.

But a major data point disrupted the trend in the early 1960s with the introduction of the rear-engine design. The technique would spell the end of the front-engine roadsters within just a few short years. Domination of the new design at Indy was quick to come after Englishman Jimmy Clark brought the first rear-engine machine home to the checkered flag in 1965. It's been that way ever since, with the powerplant sitting behind the driver.

The sheet metal changes that have evolved in the 1970s and later have resulted in the Indy Cars looking more like the Air Force F-15 fighter than something that was originally intended to run on four wheels. With the

**Right:** *As there were no electric starters for the early Indy Cars, strong crewmen had to bring the massive engines to life. Broken arms weren't uncommon, as the starter cranks sometimes snapped back on the unsuspecting crewmen.*

advent of more powerful engines, the obvious advantages of a number of aerodynamic devices such as spoilers and front air dams became obvious. The new air-cheating concepts were typified by the 1972 winning car of Mark Donohue.

But the biggest change in these cars in the modern era was the addition of wings on both the front part of the body and on the rear of the car. Wings became standard

**Left:** *One of the most popular and photographed cars that ever ran at Indy is the number 32 Marmon Wasp, driven by Ray Harroun, that won the first Indy 500 in 1911. It is on display at the Indy 500 Museum located in the infield.*

**Right:** *The technologically advanced Belond Exhaust Special won both the 1957 and 1958 races, with Sam Hanks and Jimmy Bryan doing the driving. The car featured a laid-over engine for a lower profile and center of gravity.*

**Right:** *Possibly the most famous cars ever to run at Indy were Lew Welch's NOVI Specials. Their distinctive screaming engines could be heard for miles. Even with their great power and top-notch drivers, Lady Luck never gave them a checkered flag.*

**Right:** *The late great Mark Donohue won the 1972 Indy 500 in this car. It sported huge front and rear wings that stuck the car to the track like never before. With 900 horsepower under the hood, it took aerodynamic devices like this to keep the car planted.*

**Right:** *As aerodynamic technology advanced into the 1980s, the wings became smaller and adjustable, as demonstrated on the 1980 winning Chaparral of Johnny Rutherford.*

equipment in the early 1970s, and they have stayed in place with the current Indy machines. At speeds approaching those of an aircraft, there is a tendency for these machines to take to the air. The airfoil shape of the wing is placed upside-down on these cars to produce the opposite effect of lift. The large rear wing pushes hundreds of pounds of force straight down on the rear tires. The smaller wings do the same for the front wheels, aiding greatly in the handling of the cars.

During the 1970s, the speeds kept pushing higher and higher, with the downforce provided by the wings holding the cars onto the track like glue. It would be just a matter of time until one of the teams got the cars dialed in for a qualifying run at over 200 miles per hour. This amazing-for-the-time feat would be accomplished by Tom Sneva in 1978 with a one-lap record of 203.620 miles per hour and a four-lap average of 202.156.

Even more recently, a strange phenomenon known as "ground effects" kicked the speeds of the cars up again. The wings had done their job on the top side of the car, so the designers looked to the bottom of the car for further development. A flat bottom car did nothing to increase the speed of the car, but if the bottom were shaped like a venturi, then the air moving under the car when it was at speed would be accelerated, leaving the back of the car at a faster speed.

The process actually created a low pressure area and pulled the car down to the track. The beauty of the technique was that the design didn't experience any aerodynamic drag which would have slowed the car down, like the top wings did. To be totally effective, though, there had to be a way to retain the air flow under the car. This was accomplished by so-called "skirts" that hung low on both sides of the car.

**Left:** *Wings of the 1990s feature lighter-weight composite materials and a continued down-sizing to slow the cornering speeds of these cars. The rear wing acts like an inverted airfoil, producing hundreds of pounds of downforce on the rear tires. This particular wing is on the 1991 Bill Cosby-sponsored car of the first black driver ever to run the Indy 500, Willie T. Ribbs.*

**Left:** *The rectangular side pods on an Indy Car direct air over the rear tires to improve aerodynamics. The sides of the cars hang low to contain the air underneath, producing the phenomenon known as ground effects. The suction effect holds the car to the track and is responsible for vastly increased cornering speeds at Indy.*

Many people have a real misconception about what a modern Indy Car actually is. Many think of it as a large racing machine, but really these machines are quite small, weighing in at only about 1500 pounds including the driver and fuel. Compare this with the average American sedan, which could weigh more than twice that figure.

These modern Indy Cars are about 15 feet long and are an amazing six-and-a-half feet wide. Superwide and completely treadless tires make these machines strictly dry track performers. The design rules for these low-slung machines are very simple – make them capable of straightaway speeds of up to 250 miles per hour and left hand turns only. This last rule works at Indy but not necessarily at other races, however. A number of CART (Championship Auto Racing Teams) races are run on road courses such as Mid-Ohio and Watkins Glen, where it's necessary to make both left and right turns.

Granted, modern Indy Cars are built to run fast in the forward direction, but there are times when great changes in velocity do happen to these cars. There are accidents, and designers have taken this into consideration. The hitting of a wall at speeds in the 200 mile per hour range really tests the structure of a car. Space age materials play heavily in the construction and safety aspects of modern Indy cars. Exotic carbon-carbon materials are used in the bodywork, wings, and other parts of the car. These fragile-looking machines can take amazing hits and still protect the driver. The cars are designed to "fold up" under impact and absorb the force of the impact. In recent years, the front portions of the cars have been strengthened in order to protect the feet and lower legs of the driver in a straight-on impact. Indy Car designers are always looking for ways to protect the drivers of these cars, but at these speeds it's almost impossible to guarantee total protection.

A prime manufacturer of Indy Cars is the Penske Corporation, which has built cars for the Penske Race Team and other customers for many years. The cars have won many races, both at Indy and in other Indy Car races. Lola Cars of England is also a big provider of cars for the big race, and has a lot of the front-line teams as customers. The only pure American-built Indy Car in recent years was built by Truesports Racing.

The danger of fire has always been one of the great fears in these cars, or for that matter in any type of racing machine. In earlier years, Indy Cars carried full fuel loads, which in the case of a crash presented the chance for a serious fire problem. To lessen that possibility, though, Indy Cars today are allowed to carry only 40 gallons at a time.

Everywhere you look on the modern Indy Car, it's easy to see that safety was one of the prime considerations of the design. Over the driver's head is a "rollover loop" which protects the driver's head in case of rear-end impact and offers some protection should the car flip over onto its top.

The driver is also protected on all sides by the "tub" where he sits. Constructed of a super-strong carbon fiber material, the tub can accept great impacts and still remain intact. The tires and wheels are also designed to absorb impacts. The radiators on each side of the car are filled with water and take up a lot of the impact if a crash should occur.

An interesting fact about the Indy 500 is that there is a restriction on how much total fuel is allotted for the race. The total amount allowed is 280 gallons of methanol, which requires the cars to get at least 1.8 miles per gallon. A greater gas consumption means the car just isn't going to make the distance, which makes it very difficult to win the race. Modern computers in the Indy pits monitor fuel consumption during the race. If too much fuel is being used early in the race, the driver must ease back on the engine throttle and conserve fuel.

Huge advances have also been made in horsepower capabilities to push these Indy Cars to their tremendous

**Bottom:** *You don't sit in an Indy Car – you lie in it. The driver is enclosed in a carbon-fiber "tub" which helps to protect him in an impact. The steering wheel must be removed to get the driver in and out of the car.*

**Bottom:** *The cockpit is only as wide as the driver in today's sleek Indy Cars. The seat is molded to fit the contours of a specific driver's body, thus eliminating potential driver changes.*

speeds. And even though the designs of today's engines are incredibly exotic, they still carry a multi-valve heritage all the way back to the four valves per cylinder and overhead cam design introduced by the 1913 Peugeot. That design has been carried forward in the design of a number of current Indy Car powerplants.

The engines have evolved over the years with the current V-8 design replacing the straight-eight, four cylinder and in-line six cylinder powerplants of earlier years. In the 1990s at Indy, a majority of the powerplants are V-8 configurations equipped with a crankshaft that drives all the accessory equipment.

The engines are about 161 cubic inches in displacement and have what is called a dry sump oiling system, where the oil is carried in a separate reservoir and not in the crankcase as is usually the case. A gear train or toothed belt drives the electronic systems, and an onboard computer controls the ignition, fuel feed and other important engine functions.

To greatly increase the power of the engines, they are turbo supercharged, which can practically double the horsepower of these powerplants. Loss of a turbo on an Indy Car, a common failure, will quickly put the car on the sidelines.

The current king of the Indy engine camp is the Chevy Ilmor which has dominated the race in recent years. The engine first appeared at Indy in 1986, and is capable of running at an unbelievable 12,000 revolutions per minute. It features four camshafts and 32 valves and is the state-of-the-art in engines for these cars.

One advantage of the highly successful Ilmor engine is that it is not available to all Indy Car teams. Lack of this engine under the hood of an Indy Car presents a real problem if one is interested in running up front. Research continues in a number of other camps, but for the present time, this is still the king engine of Indy.

The Chevy engine replaced the famous Cosworth engine which had dominated Indy in the same manner in earlier years. In fact, Cosworth-powered machines won ten straight Indy 500s during the 1970s and 1980s. Even today, there are still Cosworth-powered machines at the Brickyard, but they stand little chance against the

**Right:** *The Ilmor Chevrolet engine has been the kingpin at Indy from the late 1930s to date. The engine overtook the Cosworth, which had dominated Indy during the late 1970s and early 1980s.*

**Right:** *The Buick V6 engine is mostly run at Indy due to the fact that it is given extra turbo boost pressure to make it competitive with the V8 engines. It is fast but fragile, with many of the engines breaking in the early stages of the race. Its crowning glory occurred in 1985, when Pancho Carter sat on the pole with stock block Buick powerplant.*

powerful Chevys, although there is currently a new Cosworth powerplant in the works.

Over the years, so-called stock block engines have also competed at Indy. During recent years, the V-6 Buick engine has done well during the month of May. Since it is based on a stock engine, the Indy 500 gives the engine an advantage in the form of greater boost pressure than the pure racing engines offer. This makes the Buick very competitive at Indy and a number of cars equipped with these engines have appeared and run well. The engine is rated at an impressive 800 horsepower. Indy, by the way, is the only track that affords the boost advantage to the Buick, and therefore is one of the few tracks on which it runs in any numbers.

Also, during the 1980s, there was a Chevy V-6 program at Indy which was headed by the A.J. Foyt teams. And even earlier, there were a number of 390 cubic inch V-8 stock block Chevy powerplants that ran at Indy. The robust-sounding powerplants, though, never achieved any great success.

The Judd engine is a new powerplant effort that has shown real promise. The engine isn't yet competitive with the front runners, but its 730 horsepower capability at 12,600 rpm could well put it in front of the field at Indy in the years to come.

Finally, there has also been a foreign effort over the past several years at Indy, with the Italian automotive manufacturer Alfa Romeo fielding a car for the 1991 Indy 500 driven by Danny Sullivan. The effort follows a long involvement of European companies with cars entered in the big race.

A discussion of engines at the Indy 500 must make mention of the creation of Andy Granatelli during the late 1960s. The uniqueness of the car was its turbine powerplant, which whined around the track at amazing speeds. The turbine cars set track records at Indy, and came close to winning in 1968. But Indy officials wanted to keep the sound of Indy as the sound of a screaming internal combustion engine, and as a result, the turbines were legislated out of competition. They were, however, an interesting page of Indy engine history.

The stars of the great race are, of course, the sleek

**Right:** *The Judd engine program was developed and implemented by the Truesports Indy Car team. The engine has showed promise at times, but has never reached its expected potential.*

**Above right:** *Foreign engine development has long been a part of the Indy 500. The trend continues today with the Italian Alfa Romeo engine that has run for several years at Indy.*

**Right:** *The quiet "whoosh" of Andy Granatelli's turbine-powered Indy machines came close to revolutionizing the sport. One of the turbine cars sat on the pole in 1968 and nearly won the race that year and the following year. The turbine cars broke late while leading in both of those races. The fact that the turbine cars didn't have the characteristic throaty roar of a piston engine contributed to their demise.*

**Below:** *One of the prizes to the winner of the Indy 500 is a pace car. Here, 1990 winner Arie Luyendyk takes the victory lap in the Chevy Beretta pace car.*

racing machines that put on the great show for the world. But there is one other vehicle each year that isn't in the race, but which receives a huge national exposure because of its association with Indy. It is the pace car that is selected each year to pace the race. There has always been a pace car at Indy, and through the years it has become an integral part of the race. For the car company selected for the pace car, the honor is an advertising bonanza with the pace car appearing in national magazines and newspapers.

There are several criteria used by the track in selecting the pacer for a particular year. Usually, Indy likes to select an exciting new design or one that would seem to gain public acceptance. Cars like the 1955 Chevy, the 1964 Mustang and the 1967 Camaro are good examples of that criteria. Undoubtedly, much of their success can be attributed to their Indy 500 exposure.

Occasionally, cars are chosen as pace cars because of advanced technology, such as the 1949 Oldsmobile, which introduced a high compression engine, or the 1976

**Below:** *The prototype Dodge Viper was selected as the Indy Pace Car for 1991 after public outcry over the selection of the foreign-built Dodge Stealth forced Speedway officials to make a change.*

Buick with its turbocharged V-6 powerplant. The 1985 Fiero was also an interesting choice, with its innovative use of plastics.

The actual Indy pace cars (the two or three that do the actual pacing duties) are usually modified in order to be able to meet the speed requirements for the pacing duties. But that's not always the case, as several recent pace cars were hot enough in stock trim to do the job. In most cases, though, a hopped-up engine and special suspension are included for the on-track cars.

In addition to the actual pacers, there are also a number of look-alike cars that are used for official duties. Many of these replica pace cars are also produced and are available from dealers complete with all the Indy 500 identification. The cars have become very collectable in the 1990s.

One of the prizes the winner receives is one of those pace cars. Many of the recent winners have kept those cars as prized mementos of winning the biggest race of them all.

# THE DRIVERS

Running in the Indy 500 is the ultimate thrill for a race driver. Over the years, many careers have been made at the Brickyard, while many others have experienced great frustration. For some, success has come easily and they have become household names while for others, the track seems to have a jinx on them.

As tough as it is to win the great race, there have been three drivers who have won it four times: A.J. Foyt (1961, 1964, 1967 and 1977), Al Unser, Sr. (1970, 1971, 1978 and 1987), and Rick Mears (1979, 1984, 1988 and 1991). Notice the longevity of these careers, with Foyt and Unser's first and last victories at Indy occurring 16 and 17 years apart, a measure of their long-term driving excellence. The odds are against winning two races in a row, but the feat has been accomplished by four different drivers – Wilbur Shaw (1939 and 1940), Mauri Rose (1947 and 1948), Bill Vukovich (1953 and 1954) and Al Unser, Sr. (1970 and 1971).

Experience seems to be one of the keys to driving the challenging low-banked Indy track. There are many drivers in their late forties and even their fifties that continue to drive at Indy and show success. It seems that some careers will just never end at the Brickyard. Take A.J. Foyt, for example, who has been in 34 consecutive races. Other fixtures in the recent era have included Johnny Rutherford, Mario Andretti, Gordon Johncock, Al Unser, Sr., Bobby Unser, and Lloyd Ruby.

Following are brief biographies of some of the "Kings of Indy," reviewing their accomplishments at the Brickyard.

## A.J. FOYT

Anthony Joseph Foyt, Jr., is the unofficial master of Indy. His long career of excellence includes four victories and 11 other top ten finishes.

A.J. was a brash 23-year-old when he first ran at Indy in 1958. He started in 12th place and ended up 16th. That first year, A.J. qualified his Dean Van Lines Special at just over 143 miles per hour. By 1991, his qualifying figure was over 220 miles per hour.

A.J.'s first Indy win came in only his fourth year, when he led the race for 71 of the 200 laps. Three years later, in his Sheraton-Thompson Special, A.J. won again and totally monopolized the race, leading 146 laps. He might have won the next year in 1965 when he was the leading qualifier and led for much of the race, but a broken gearbox put him on the sidelines. He won the race for a third time in 1967.

In the next decade, there were times when A.J. could have won the race, though he never did. He was the fast qualifier two times and led over 200 laps during the time period, but mechanical failures kept putting him out of competition.

**Above:** *Without a doubt, A.J. Foyt is Mr. Indy 500. He is the first four-time winner, and has come close a number of other times.*

**Above right:** *Foyt in 1984.*

**Right:** *In 1989, A.J. started tenth and finished fifth.*

That fourth win would finally come in 1977 for the Gilmore Racing Team and A.J. when he took the checkered flag after starting fourth. Since then, A.J. has tried as hard as he can to find that elusive fifth Indy win.

At 56 years of age in 1991, A.J. was still hanging in there, giving it his best shot. Even after a major crash in 1990 that caused serious injuries to his legs, A.J. went through a rehabilitation that amazed even his doctors. The huge rounds of applause the old Indy master received from his admiring fans in 1991 every time he appeared are testament that he is probably Indy's most popular driver ever, bar none.

GILMORE
14
Copenhagen

## RICK MEARS

The best-known Indy driver of recent years is undoubtedly Rick Mears, one of three four-time winners.

Rick came to Indy Cars from the wild and crazy world of off-road racing. He finished third in the Pike's Peak Hill Climb in 1973. In 1976, he ran his first Indy Car race, an excellent eighth place finish at the California 500.

His first try at Indy came in 1977, and he didn't make it. But that wouldn't happen again. Besides his brilliant performances during the races themselves, Rick is an absolute master at qualifying on the old track. He's been on the pole for the great race six times. Considering the high level of competition at the Brickyard, his accomplishment is almost unbelievable. The combination of Mears's excellent driving skills and the meticulous car preparation of the Penske Racing Team is a real winner.

But besides those top qualifying efforts, Rick has also been second-quick once, third-quick three times and fourth-quick one time. His "worst" efforts were two tenth-quick qualifications. Rick led the race for at least one lap from 1979 through 1984, a new Indy record which may well never be broken. His race performances were equally impressive during that time, with two firsts, a second, a third, and a fifth. The 1982 race, where he lost to Gordon Johncock by a matter of inches, could have been an additional victory with a little luck.

Rick Mears wears his Indy titles well. He is known as a real class act.

**Left:** *Rick Mears became the most recent member of the exclusive four-wins club with his 1991 victory. Having access to Roger Penske's research and development capabilities, his equipment is always top-notch.*

**Left:** *One of the pleasant required duties for a driver is posing with the winning car the day after the race. Here, Rick Mears does the honors following his third Indy victory in 1988.*

**Right:** *Rick Mears has won a record six pole positions. This photo shows his fifth, in 1989. Also on that front row were Penske teammates Danny Sullivan and Emerson Fittipaldi.*

**Left:** *The second member of the four-victories club was Al Unser, Sr., whose wins came in 1970, 1971, 1978 and 1987. He was the oldest driver ever to win the race when he took the 1987 event. He also was the fast qualifier for the 1970 race.*

**Above right:** *Here's Al at speed during the 1972 race in the Viceroy Special. He finished fourth in this race after back-to-back victories in 1970 and 1971.*

**Right:** *In a rags-to-riches story, Al came to the Speedway without a ride in 1987. After Danny Ongias was injured in a crash, Al picked up the ride in a car that had been used earlier as a show car, and went on to win his fourth event.*

## AL UNSER, SR.

An Indy legend of major proportions, Al Unser, Sr. has run at Indy during four decades and shows four victories at the great track. Before coming to Indy, he was a part of the Unser family monopoly at the Pikes Peak Hill Climb. He also was an excellent performer in midgets and sprint cars.

It was obvious that Al Unser was something special when he finished ninth in his first Indy in 1965. Two years later, he was second and looking for victory; it came in the 1970s with two straight victories, in 1970 and 1971. The 1970 win was an overpowering effort, with Al in the lead in 190 of the 200 laps. He was also the quick qualifier that year. The following year, he led in 103 laps for his second straight win.

Had it not been for a broken piston, Al could well have been in contention for his third win in 1973. He led for 18 laps that year before being sidelined. Close again in 1977 when he was fourth quick, he led for 17 laps, and finished a close third.

Running the revolutionary Chaparral car in 1978, Al took Indy win number three leading the race for 121 laps. He also won the 500-mile races at Pocono and Ontario speedways.

During the 1980s, he was close a number of other times with a fifth in 1982, a second in 1983 when he led for 61 rotations, a third in 1984 and a fourth in 1985. Then came the amazing story in 1987, when he came to the speedway without a ride, stepped into a Penske backup car and led 18 laps for the victory. He led the race on three different occasions the following year and again finished third.

Al Unser is truly an Indy legend, with 625 laps led during his 25 races. It's an Indy record that will stand for a long time. He did not run in the 1991 race, but don't count out this savvy Indy veteran. He's possibly the best ever to drive at the old Brickyard.

25
Cummins

**Left:** *What do three-time Indy 500 winners do after they retire? Many of them are asked to drive the pace car, like Bobby Unser, shown here with the 1989 Pontiac Trans Am Pace Car. Bobby, the older brother of four-time winner Al, also maintains a close association with Indy as a TV color announcer.*

**Below:** *Bobby's second win at Indy came in 1975 behind the wheel of this robin's-egg blue Jorgensen Eagle built by Dan Gurney. Bobby also won the race in 1968 and in 1981, which gives him a win in three different decades.*

## BOBBY UNSER

Regarded as one of the all-time great Indy 500 performers, Bobby Unser – older brother of Al – is a three-time winner of the race.

Bobby started his racing career with stock cars in 1949 when he was only 15, then moved to midgets and sprint cars. Like all the other racing Unsers, he ran the Pikes Peak Hill Climb a number of times and won the Championship Car Division in 1956.

He arrived at Indy in 1963, but didn't do much until 1968, when he set a track record in qualifying. He led the final eight laps of that race to take his first Indy victory.

Bobby set another qualifying record in 1972 of 196.678 miles per hour, but a broken motor disappointed him in the race and he finished 13th. He was close again in 1973 when he was the quick qualifier and led the first 39 laps until his engine quit. In 1974, he finished second, only 22 seconds behind winner Johnny Rutherford. In 1975, Bobby became the ninth two-time winner of Indy. His final win came in 1981.

In recent years, Bobby has served as a color commentator on TV talking about what he knows best – the sport of Indy Car racing.

## JOHNNY RUTHERFORD

Lone Star J.R.: That's how most Indy 500 fans know Johnny Rutherford, a real Indy legend. He's been involved with Indy for almost three decades and is a three-time winner. He may well have run his last Indy, but he still stands tall as one of the finest ever to wheel those bricks.

His Indy initiation was in 1963, and it would be eleven long years before he would win his first. The 1974 win was especially sweet, since Johnny started in 25th position and sped through the field for the victory. Amazingly, he would lead 122 of the 200 laps even with his far-back start.

Two years later, J.R. did it again, taking all the marbles as the quick qualifier and the victor. The year between those two wins could have made it three in a row, but he finished a close second in 1975. He was really on a roll.

Four years after his second win, in 1980, he was overpowering again with fast time and the victory. He led 118 laps in that victory, more than a third of the 296 Indy laps he has led in his career.

Unfortunately for J.R., he never got close to win number four during the 1980s. But his driving skills kept him in the top ten with a sixth and two eighths as his final best efforts. His last Indy 500 was driven in 1988, but you just know that if he could get a good ride, he'd love to give it one more shot for number four.

Like Foyt, Andretti and the Unsers, Rutherford is one of the last of the old school to come up through the small track racing of sprint cars and midgets and make it big at Indy.

**Below:** *"Lone Star J.R.", a.k.a. Johnny Rutherford, has long used the Lone Star symbol from the state of Texas as a trademark on his helmet. Success at the Brickyard is another of his trademarks, with three victories in 1974, 1976 and 1980.*

**Bottom:** *During his later years at Indy, rides have become harder to acquire for J.R. Here he drives one of fellow Texan A.J. Foyt's backup cars in 1984. The car qualified 30th and finished 22nd.*

**Below:** *Throughout his career, Mario Andretti has been one of the most consistent and fastest drivers ever to grace Indy. Yet, he has only been able to accomplish one victory at the Brickyard, in 1969.*

**Bottom:** *Mario holds the pole position on the parade lap of the 1987 race. It was one of three poles he has held during his lengthy Indy career, the other two being in 1966 and 1967.*

## MARIO ANDRETTI

Mario Andretti's name is associated with many types of racing. The Nazareth, Pennsylvania driver has done it all with sprint cars, NASCAR stock cars, modifieds, Formula 1 cars (he was the World Champion in 1978), and finally, Indy Cars for which he's probably best known.

Mario has been in the Indy starting field for 26 straight years. But the great race has been the ultimate in frustration for one of the world's great race drivers.

The first and only win for Mario came in only his fifth year (1969). It looked like it might have come sooner, since he set fast time in both 1966 and 1967. He would finish 18th and 30th those two years after leading 16 laps in the 1966 race.

Always a great qualifier at Indy, Mario was out of the fastest ten cars only three times. But it just hasn't translated to success during the race. He's come so close on numerous occasions, but car failure at the most inopportune times has usually put him out of the running.

In 1980, he led the race for ten laps and then the engine seized. Four years later, he led the race for 29 laps and then was taken out in a crash on the 153rd lap. His most disappointing Indy, though, had to be in 1987 when he looked like a certain winner after leading for many laps, until the engine gave up.

He's in his fifties now, but he's surely going to stick around as long as he can and hope to win the biggest race in the world a second time before he retires.

**Below:** *Brazilian-born Emerson Fittipaldi added an international flavor to the great race when he first appeared at the track in 1984. Once he combined his superb driving skills with the resources of the Penske team, Emmo became a threat to win every time out.*

## EMERSON FITTIPALDI

A relative newcomer to Indy, this driver brought international credentials to the speedway. He's run just about everything, including motorcycles, go-karts, sedans and Formula 1 cars.

As a Formula 1 driver, he won the World Championship in 1972 when he won five races and beat out Jackie Stewart for the title. He ran in Formula 1 until 1981, when he dropped out of racing completely for three years.

In 1984, he started his Indy Car career and qualified for the 1984 Indy 500. It wasn't a great beginning, though, as he qualified 23rd and finished a next-to-last 32nd with engine failure.

It wouldn't take long for Emerson to establish himself as a threat, though, as he qualified fifth the following year and finished 13th. Then came three great years (1988 through 1990) when he qualified eighth, third and first, and finished second, first, and third in the race. With a little luck, that third place could well have been a second victory, as he had led the first 92 laps and 128 laps overall. A broken gearbox in the 1991 race put him on the sidelines on the 171st lap.

A teammate of Rick Mears on the Penske team, Emerson is part of one of the most formidable teams ever to challenge Indy.

**Bottom:** *Fittipaldi's first Indy victory came in 1989. His second place finish in 1988 and third place in 1990 gave him an impressive string of high finishes.*

## ARIE LUYENDYK

Like Emerson Fittipaldi, Arie Luyendyk ran Formula 1 cars to groom his open wheel skills before coming to Indy in 1985. That rear engine experience paid big dividends for the Dutchman who won the 1990 Indy 500.

He made his Indy Car debut in 1984, but it wasn't at the old Brickyard. The following year, he qualified 20th for his first Indy. He certainly raised eyebrows that year, with a seventh place finish. A year later, an accident took him out of the race with only 12 laps to go. He finished 15th.

In 1987, Arie qualified well with a seventh place effort, but a suspension failure dropped him to an 18th place finish. The following year, he moved up to a sixth place qualification effort and a tenth place finish. It was like a big step backwards in 1989 when he qualified 15th and finished 21st with a blown engine.

Was he always going to be an also-ran? Hardly. That all ended in 1990 when he was the third fast qualifier with a 223.304 mile per hour clocking. Then, he surprised everybody as he led 37 laps and took the checkered at a 185.981 mile per hour average.

Things will never be the same again for Arie. He's now won the Indy 500, which sets him apart in a very select group of drivers. Besides taking home $1 million for the victory, he became a part of the legend called Indy.

**Bottom left:** *Arie Luyendyk honed his driving skills in the European Formula ranks, which gave him a multitude of experience to deal with driving high-powered rear engine machines much like the Indy Cars.*

**Bottom:** *Luyendyk surprised everyone in 1990 by winning from his outside front row starting position. He can no longer be overlooked as a contender for the Indy crown.*

**Below:** *Bobby Rahal has proven himself to be one of the Indy 500's most consistent performers.*

**Bottom:** *Besides his emotional win in 1986 for cancer-stricken owner Jim Trueman, Bobby has also had a number of other high Indy finishes.*

## BOBBY RAHAL

Bobby Rahal looks like a college professor, and indeed he does hold a degree in history from Denison University. He brought an outstanding road race background to Indy Car racing. He ran with the Sports Car Club of America (SCCA) during the 1970s and was the national champion in several classes during that time period. In 1980, he competed in the Can-Am series and was fifth in the points. In 1981, he won his class in the 24 Hours of Daytona and was third in the World Endurance Driver's Championship.

He qualified for his first Indy in 1982 and finished an outstanding 11th. He would go on that year to win his first Indy Car race at Cleveland. In 1983, he actually led 15 laps in only his second Indy, but had to retire with a hole in the radiator. An engine problem cut short an excellent third-quick qualifying effort in 1985, when he led for 14 laps.

In 1986, Bobby led the race for 58 laps, but most importantly, he led the 200th lap, giving him the victory. During the late 1980s, Bobby was involved in the development of the Judd racing engine and was not as competitive at Indy as he would have liked to have been.

But in 1989, he joined the Kraco Race Team and again was in the hunt. He came very close to winning in 1990 when he finished second behind Arie Luyendyk. An engine failure in the 1991 race dropped him down to 19th position.

But rest assured, Bobby Rahal could well become one of the legends of Indy by the time he's finally done at the Brickyard. He is already recognized as one of the best drivers ever to come to Indy.

**Below:** *A three-decade Indy performer, Tom Sneva had his first and only win in 1983. He has also had three poles, two other front row starts, and six top ten finishes including three in second place.*

**Bottom:** *Sneva is known as "The Gas Man" for two reasons: he was the first qualifier to run a 200 mile per hour lap, and he drove the Texaco Star car for a number of years.*

## TOM SNEVA

They call him "The Gas Man" – and when Tom Sneva's accomplishments are examined, it's easy to understand why that's an appropriate label. Running in three decades at Indy, Sneva brought excellent credentials to the Brickyard with experience in stock cars, modifieds and sprint cars. He ran his first Indy Car race in 1971, and debuted at the speedway in 1973. He passed his driver's test that year, but made no qualifying attempt.

He made an outstanding showing in 1974 with a ninth quick qualification run and finished 20th. He finished sixth in 1976, and then came two top qualifying efforts the following two years. Those feats are how Tom Sneva is best remembered at Indy.

In 1977, Tom ran the first 200 mile per hour lap, and then the following year set the four-lap record of 202.156 miles per hour. He continued his top-notch qualifying accomplishments with a second-quick effort in 1979, and top runs in 1981 and 1984.

Luck on race day finally went Tom's way in 1983, when he started fourth and came home first to the checkered flag. He led 98 laps in that convincing victory. He would lead 31 laps of the 1984 Indy, but would fall out after 168 laps with mechanical problems. In all, Tom has led 208 laps at the Brickyard.

**Below:** *His Hollywood good looks and style have made Danny Sullivan a real favorite of the Indy race fans. He's also got great driving skills to go with those looks.*

**Bottom:** *Danny spent one year (1984) driving the Domino's Pizza Hot One before moving over to the Penske Miller American Team, where he has enjoyed his greatest success. His first and only victory came in 1985.*

## DANNY SULLIVAN

Dashing good looks and a skillful style on the track have made Danny Sullivan a favorite at Indy. He brought an excellent road race career to Indy, with experience in European Formula 1 racing and the Can-Am series.

After two 17th-quick qualifying efforts in 1982 and 1984, Danny amazed the racing world by winning the 1985 race, leading the race for 67 laps. But the memorable happening in that race was the famous 360-degree spin that Sullivan negotiated, and recovered from, to go to victory in his first year with the Penske team.

It looked good for Danny again in 1986 when he sat in the middle of the front row with a second-quick qualifying effort. He broke the track record at the time with a 215.382 mile per hour run. He finished ninth that year.

Danny ran strong in 1987, starting in 16th position but leading the race for four laps before engine failure on the 160th lap. In 1988, he set a new track record of 216.214 miles per hour, a mark which was broken 39 minutes later by Penske teammate Rick Mears. He led 91 of the first 94 laps, but again luck was against him as he was put out of the race with a first turn crash, relegating him to a 23rd place finish.

In 1991, Sullivan left the Penske team and drove for the Alfa Romeo team. His Indy performance was a disappointing tenth place finish, but look for more things at the Brickyard from this driver before his career ends.

## UP-AND-COMING DRIVERS OF THE 1990S

The Indy stars of the 1990s and beyond are learning the trade today, and there are some familiar family names, such as Andretti and Unser, that will continue to be prominent. Michael Andretti and Al Unser, Jr. have both already come close to winning the big race, and both will undoubtedly win Indy before their careers are over.

Michael ran his first Indy in 1984, starting fourth and finishing fifth. In 1988, he was fourth, and in 1991 he almost won the race, finishing only 3.1 seconds behind Rick Mears in a duel to the checkered flag. He was also a dominant force in the 1991 CART Indy Car title race, winning the championship over Bobby Rahal.

Al Unser, Jr. has also been close to victory. In the 1989 race, he finished second to Emerson Fittipaldi, and in both 1990 and 1991 he finished fourth.

John Andretti, nephew of Mario, and Jeff Andretti, Mario's younger son, are also promising young drivers who will carry on the great Andretti tradition in racing.

**Below:** *Clearly born with his famous dad's driving skills, Michael Andretti appeared on the Indy scene in 1984 and finished fifth. He was Co-Rookie of the Year.*

**Below:** *Son of four-time Indy winner Al Unser and nephew of three-time winner Bobby Unser, Al Unser, Jr. is a good bet to add his name to the Borg-Warner winners' trophy.*

**Bottom left:** *Nephew of Mario and cousin of Michael, John Andretti carries on the legacy of the family name. His father Aldo was also an excellent driver before being seriously injured in a sprint car crash.*

**Bottom middle:** *Michael Andretti has already had five top ten finishes, including second place in 1991. He has also qualified in the top ten six times out of eight races.*

**Bottom right:** *Al Unser, Jr. has been in the top ten qualifiers six times and has finished in the top ten six times in only nine years. He was leading in the 1989 race when he touched wheels with Emerson Fittipaldi and spun into the wall on lap 198.*

# SPEED AND COMPETITION

Speed and competition is the name of the game at the Brickyard during the month of May. But there is a lot more to competition than just the cars racing against each other on the huge track on race day.

Carrying almost as much importance as the racing itself are the activities in the pits. Getting in and out of the pits as quickly as possible can well mean the difference between victory and defeat at Indy. The precision of the top Indy Car pit crews is poetry in motion. And considering that a car can cover the length of a football field every second, the importance of getting in and out of the pits as quickly as possible is easily understood. The crews practice throughout the month of May to get that pit precision down to a fine art.

But the blazing speeds on the track are what the fans really relish. That's why the qualifications at Indy enjoy a huge popularity, the opening day of which usually draws crowds of over a quarter of a million. In the qualifications, a single car is pitted against the track. Nothing but speed is being considered, and anything to acquire that extra tenth of a mile per hour is attempted. It can mean the difference between making and missing the great race.

Speaking of those qualification speeds, they have continued to increase through the decades even though there have been rules to slow the cars down for safety reasons. In the early years, those old cloth-helmeted gladiators qualified at 88 miles per hour, an amazing speed at the time. By the 1930s, the qualification speeds had pushed into the high 120s, which "experts" at the time felt was the maximum for the low-banked oval. During the 1950s, there was a saying around the track that there would never be a 150 mile per hour lap traversed at the Brickyard. That sacred speed milestone, though, would fall in 1962 when Parnelli Jones would turn a 150.370 mph qualification run.

The next two decades would continue an upward spiral of speed at Indy. During the 1960s and 1970s, the qualification speed increase would average well over four miles per hour for each year. When the first turbo supercharger was introduced in the late 1960s, the speeds and performance increased as never before. Even though the allowable engine displacement had dropped from about 270 cubic inches in the 1950s to the middle 250s in the 1960s, and finally to the present 161 cubic inches in the late 1970s, the speeds kept climbing and climbing at Indy.

During the early 1970s, the speeds were positively skyrocketing. The top qualifying speed in 1971 was 179.354 by Peter Revson, and a year later, Bobby Unser was over 17 miles per hour faster. Two signficant changes were responsible for the huge improvements – wider racing tires, and the huge wings that were starting to appear on the Indy cars. This allowed the cars to be tacked down to the track like never before. Also, there were no turbo boost restrictions on the engines, so the

**Left:** *Practice and fine tuning can make the difference between making the race and going home. Thousands of dollars are at stake with each adjustment of the chassis, engine and suspension. These are sensitive machines and minute changes can make big differences on the track.*

**Above right:** *The rear cowl of Kevin Cogan's 1988 machine has been removed during a practice session in May. The crew must have come up with some of the right answers, as Kevin finished eighth in the race.*

**Right:** *Practice, practice, practice can mean success on the track – sometimes. For Mario Andretti in 1990, it meant an excellent sixth qualifying position, but engine problems caused him a 27th place finish in the race.*

MACHINISTS
FASHIONS

Havoline
Gillette
Emerson Radio
Havoline
rain-x

**Left:** *No detail is overlooked when it comes to the quest for speed during qualifications. Only the 33 fastest cars will be in the race. The crews used to polish the wheels, as one of A.J. Foyt's crewmen is doing here for the 1984 qualifications. Now, there are high-tech wheel covers which aid the aerodynamics of the car.*

**Above right:** *The AMAX Penske Chevrolet of Tony Bettenhausen is directed off the track by a pit official during a practice session prior to the first day of qualifications.*

power was available to pull the cars down the straightaways in spite of the huge drag-inducing wings.

The sanctioning United States Auto Club soon came to the conclusion that maybe the cars were running too fast at Indy. For 1974, the turbo boost pressure was lowered to 80 inches and the wings were reduced in size. Subsequently, the top speed that year dropped seven miles per hour, and didn't come back up easily thereafter. Other restrictions were initiated in 1976 as USAC tried to stay ahead of the Indy teams. It seemed as though a magic 200 mile per hour lap at Indy would never be possible.

By 1977, the speeds were on the rise again as Tom Sneva ran an awesome 198.884 mph lap, with one of those rotations actually being over the unbelievable 200 mile per hour barrier. The huge increase in performance was thanks to the new Cosworth engine, which was capable of some 900 horsepower even at the reduced boost level.

In 1979, performance-minded Indy fans were shocked

**Overleaf:** *The quest for speed at Indy often finds teams using more than one car to come up with the right combination. Here, Roberto Guerrero has his back-up 2T car on pit row in case something happens to his primary machine.*

again when a huge 30-inch drop in engine boost pressure (some 37 percent of the total that was previously available) went into effect at the Brickyard. Indy teams said at the time that they would be lucky to run 180 miles per hour at the old track. Giving up more than 200 horsepower, it appeared that the stylish Indy Cars would be reduced to mere passenger car speeds. And to make sure that everybody conformed to the rules, USAC installed a so-called "pop-off" valve on all the engines, which would not allow more than the mandated 50 inches of boost.

The teams responded with more technology, and the 1978 pole speed by Rick Mears of over 193 miles per hour showed only a seven mile per hour reduction, certainly a lot faster than anybody thought could be run under the circumstances. Improved tires and different driving techniques were the main reasons for the moderate decrease in speed even with the major power reduction.

During the early 1980s, advanced technology would strike again at Indy with a technique known as ground effects. The innovation was, in effect, a way of increasing the down force on the car. Designers had learned in the 1970s that the wings could provide a tremendous advantage in getting the cars through the Indy turns. But the problem was that the wings tended to slow down the cars on the long straightaways.

So the bottoms of the cars were examined for a possible solution. It was discovered that if the bottom of a car was shaped like a venturi, there would be an acceleration of the air moving under the car. A low pressure area would be formed, pulling the car down to the track with effectively no external aerodynamic drag.

But in order to retain the effect, it was necessary to hang "skirts" along the bottom sides of the car, building an air seal and creating a tunnel under the car. Several of the drivers and designers said the design was like having an extra hundred horsepower at Indy.

In the 1980s, USAC tried again to slow the cars down, even threatening to do away with the ground effects completely. That threat never came to pass, but the engine boost was again reduced, this time to 47 inches, about half of that which the engines were running in the early 1970s. Still, the speeds kept moving up during Indy pole day, to the delight of the fans, as designers and crews kept finding ways of keeping up the speeds.

Minor rules changes continued to be made, but the

STP
2T
BOSCH
EAGLE

STP
CART
GOODYEAR
EAGLE
GOODYEAR

**Below:** *Tom Sneva knows the importance of qualifying well at Indy. He's won three pole positions and has been in the top ten a total of ten times. His cars through the years have always been meticulously prepared.*

speeds kept moving up, with the record run to date being an unbelievable 225.575 mile per hour run by Emerson Fittipaldi. An even more amazing effort was accomplished by 1990 winner Arie Luyendyk, who ran the 500 miles at a race record speed of 185.981 miles per hour.

But the speed a car can attain during qualifications has very little to do with how it will do in the big race itself. All that speed and performance is meaningless if the car is unable to complete the 500 miles. That's why, as soon as a car has qualified for the race, crews immediately turn to building the "race day set-up." The work is based on determining how the car will perform with a full load of fuel.

After qualifications, the drivers get a feel of whether the car is "pushing" or is "loose." A car that is pushing means that it tends to keep going straight when it should be turning, which could cause it to go into the wall. A car that is too loose has a tendency to spin out. Indy crew members adjust the chassis so that there is a comfortable combination for the driver to negotiate the 500 miles.

The skill needed to drive one of these cars at Indy is immense. An Indy Car driver explained, "When you're driving at well over 200 miles per hour, you don't just look 50 feet ahead on the track. You look a quarter- or half-mile ahead. At the speeds we are running, a great distance is covered during the blink of an eye."

**Below:** *Michael Andretti's 1989 qualifying speed was 220.486 mph. He started fifth and finished fourth that year.*

With the lower power of more recent Indy Cars, Mario Andretti emphasized that there is a different way of driving the low-banked Indy track. "We run these cars pretty much flat out all the way around the track," he observed. "We used to lose as much as 20 miles per hour going through the turns, but now we're only down as little as five or six miles per hour."

Because the Indy Cars are so low-slung, a different driver's position is required in the car. And to keep the shape low to the ground, it's impossible for the driver to sit up in the car. The driver is wedged into the cockpit in a seat that has been exactly contoured to the shape of his body. The driver is actually in a banana-shaped position, peeking over a small windshield. It's not the most comfortable position in which to deal with the wild speed, suspense and split-second decisions to be made on the track.

Running the Indy 500 is a draining experience, both emotionally and physically. Completely encased in a sweltering Nomex driving suit and temperatures that can get as high as 150 degrees in the car, it's an exhausting experience requiring the drivers of the great race to be in excellent physical condition.

There's also the 2000 turns that must be negotiated through the course of the race. The centrifugal force pulls at the body and pulls particularly at the head. Top

**Below:** *Brazilian Raul Boesel brought a wealth of European road racing experience to Indy. In spite of his many victories in other types of racing, he's only had one top ten start at Indy and two top ten finishes in five races.*

**Below:** *The newly-added Valvoline speed sign provides the fans with an almost-instant reading of a car's previous lap speed. For many fans, the qualifications are the most exciting part of Indy.*

**Below:** *With cars on the track covering hundreds of feet every second, the importance of efficient pit stops is obvious. The crews practice throughout the month to make the stops as quick as possible.*

**Below right:** *Tires play an extremely important role in Indy Car performance. Each set of tires is meticulously matched and labeled for use in the race. A wrong tire can upset a car's balance and handling.*

physical condition is imperative in driving these cars.

One driver indicated that one of the big surprises for a first-timer at Indy is the difference between driving before empty and full stands. Most of the drivers experience miles per hour lost when the stands are full. "I don't know why it happens, but when you get all that color in the stands, the track just seems to get narrower," an Indy driver explained.

Other drivers have noted that with a number of cars on the track, there is a problem of turbulence, or disturbed air, which buffets the cars. Also, as the fuel is consumed, the handling characteristics of the car keep changing, causing the driver to keep adjusting his style on the track. Indy champion Emerson Fittipaldi said that with a full fuel load, the car is heavy and the steering is harder to handle.

But then again, it's all for naught if the car can't efficiently come into the pits, quickly acquire fuel, chassis adjustments and service and get back onto the track. Indy pros stress that the race is won or lost in the pits. For that reason, the procedures during pit stops are practiced over and over again. They have to be second nature for a pit stop to be carried out to perfection.

In fact, in recent years at Indy, the job of the pit crews has been recognized through a Pit Stop Contest held several days before the big race. The cars move from a standing start into a simulated pit area, then the crews come over the wall to change two tires and the cars roar out to the finish line.

There are a number of different types of pit stops. There are stops for fuel only, although no matter what the reason is for a pit stop, the fuel will usually be topped off. There is also the frequent need to change right side tires. Other times, there will be a complete tire change. What is done often depends on the situation on the track and where the particular car is in the race. Of course, the best time to make those pit stops is under a yellow flag situation.

At Indy, space is very limited in the pits and as a result, the pits are a very dangerous location. With cars screaming by crew men just inches away, complete attention to what is going on around them is paramount.

There is also the ever-present danger of fire, as fueling is carried out in the vicinity of red-hot exhaust stacks capable of starting an inferno of invisible flames. Fires have occurred in the Indy pits, but safety rules have continually been added to make them as safe as possible.

A tremendous amount of time, effort, and money goes into making and keeping a car competitive for the ultimate challenge: the race.

**Above:** *Pit boards still play an important role in relaying information to the driver. Many radios are one-way from driver to crew to avoid disturbing the driver's concentration on the track.*

**Right:** *Computers now play an important role in Indy Car race strategy. One of their most important applications is that of fuel management, telling the team when to pit and also when to conserve fuel.*

# THE RACE

The buildup to the Indy 500 usually begins on the first Saturday of the month of May, with elaborate opening ceremonies. After the usual speeches and the appearance of the Speedway High School Band, the official band of the speedway, it's time to get started "shaking down" the machines.

An interesting competition of sorts takes place that day to see which can be the first car on the track. There's no award or money of any kind, but some teams in the past have looked at it as a good way to start a month of competition. In fact, on more than one occasion, two cars have actually drag-raced down the pit row to win the honor.

The next week proceeds with lots of hard work for the Indy teams. Fine tuning, chassis set-ups, and changes in wing settings are all part of getting the cars ready for the run for the roses. The speeds that are reached during this week are big news as fans try to figure out who will be sitting on the pole the following Saturday.

Pole Day brings out as many as a quarter of a million or more fans to see this battle of man and machine against the clock. Qualifying positions are drawn and the cars are lined up for their individual attempts to become one of the chosen 33 on the grid. Down the pit road, the United States Auto Club technically checks every car to ensure that it conforms with every rule and dimension. If a car doesn't meet the specifications, it is sent back to the garage area for corrections, and eventually returns to the end of the qualification line to try again.

The suspense builds up as each car rolls out and the Indy 500 announcer drawls out "He's on it" as a car crosses the yard-of-brick starting line. The lap times are announced quickly during the run, and when there's a

**Below left:** *A.J. Foyt's team prepares for the start of the 1990 race. In 34 straight starts, Foyt has won numerous pole positions, had many top ten finishes, and was the first four-time winner.*

**Below:** *Although the drivers get 99 percent of the publicity, the skilled crews are the heart and soul of every race team. On Carburetion Day, they have their time to shine in the Miller Pit Stop Contest.*

good effort, it's quickly noted on the PA to the cheers of the fans.

The fast qualifier that first day has an added advantage of acquiring the pole position for the race. Many times, more than half of the field is filled that first day. But there are three more days of qualifications left: the following Sunday and the next weekend give the non-qualifying teams a chance to try again.

A car has three chances to make the race, but any time a car records four laps, that will be the official time for the car. When a team doesn't think that its car is going fast enough to make the race, the car will be waved off by the crew. It has to be a quick decision by the crew, even though the car may not be able to run faster or may break down the next time out. There have been occasions where a car was waved off with a time that would have made the race.

But the real suspense occurs on the final qualification day. The rules state that the fastest 33 cars make the race, so a procedure known as "bumping" takes place. The slowest car in the field is in a precarious position known as being "on the bubble." As the final few cars go

**Right:** *Next to winning the Indy 500, just making the great race is the ultimate thrill. Here, Jim Crawford looks pleased after his qualifying run for the 1991 race. The likeable Scotsman, who was using Buick power, finished a disappointing 26th.*

BUICK
PPG
Quaker State
SIMPSON
26

out on the track, the bubble team hopes and prays as the speeds are announced. The figures either bring joy or sadness to that team.

Many times, deals are made in those final hours as drivers search for a ride to make the race. Cars are bought and sold and drivers jump into unfamiliar cars to make that final run. It's all very exciting and many times runs right down to the final second before the field is set. Sometimes a car is on the track when the final gun sounds. That car is allowed to complete its qualification run and makes the race if its speed is fast enough.

After a car is qualified, Indy teams then turn to the important job of setting up their cars for race day. The set-up for running with a full load of fuel along with working on pit stop procedures now comes into play.

The Thursday before the Sunday race is called Carburetion Day. The name comes from earlier days at the Brickyard when Indy Cars used carburetors for fuel-air management responsibilities. This is another exciting day, as the whole field is given two hours of green flag time on the track for final tuning. If you can't go to the race, this is second best. A "small" crowd of only about a 100,000 shows up for this event. This is also the day that the Pit Stop Contest takes place, where the crews get their chance to shine. Most of the crowd stays in place to watch this unique competition.

But there's another attraction on the Indy grounds that is open all year around and provides the fan with a real understanding of the history of the Brickyard. The Hall of Fame Museum recalls all the great years of racing at the track with displays of more than 75 race cars. Many former winning cars are on display, including the cars from three of the first four races. The march of racing technology at Indy is clearly illustrated, as many of the winning cars through the years have been donated to the museum by their teams. It's also possible to take a bus ride around the track in order to visualize what the drivers see as they rip off the second and fourth turns and roar into the long, narrow straightaways.

On the day before the race, all the drivers assemble and receive instructions for the race itself. It's a real media event, with all the famous drivers sitting together in a block. A huge parade through the streets of Indianapolis follows. The stars of the parade are the drivers themselves, who ride in individual pace cars and wave to the immense crowds that line the streets.

In the wee hours of race day, the gates to the great track are swung open and thousands of fans race to their favorite spots. These are the fans who don't have reserved seats and are striving to find a good location for what is possibly the biggest party in the country in the Indy infield. For many, though, the party has been underway for days as many fans of racing, and just fans of having a good time, have been camped by the thousands in neighboring fields.

The heart of the party at Indy has long taken place in a location on the first turn known as the "Snake Pit." It's debatable how much of the race the occupants of this area actually see, but then again, the Indy 500 is a happening of unprecedented proportions, and people enjoy the race in different ways. In recent years, the area has been reduced in size as new grandstands are constructed.

An army of employees runs this giant speed plant.

**Left:** *One of the most famous symbols of victory in sports today is the Borg-Warner Trophy, awarded to the victor of the Indy 500 each year. A likeness of each winning driver's profile becomes part of the trophy to immortalize the accomplishment.*

**Right:** *Hollywood stars have had a long association with the Brickyard. For many years, a popular Hollywood starlet would greet the winner with a kiss in Victory Lane. In recent years, stars such as James Garner have participated in the 500 Festival Parade.*

**Below:** *Pre-race pageantry on race morning is led by the Purdue University All-American Marching Band. The band has been a fixture at the race for many years.*

Some 2400 members of the Safety Patrol control the traffic and the thousand other details that have to be undertaken to run this extravaganza. The staff also includes 250 doctors and 1100 concession workers.

The great event brings celebrities from all over the world, some of whom are intimately involved with the racing itself. Movie actor Paul Newman, for example, is a part owner of the Andretti Race Team and can be seen on pit row much of the time. James Garner, a long-time racing fan and a pace car driver several times, is also a frequent visitor to the track.

The excitement starts to build as race time nears. The cars are moved to the starting grid on the track. Then the procedure before the race starts, a ceremony which is pure Americana and brings a shiver to even the non-racing fan. The sound of the song "Back Home Again in Indiana," which has been sung in recent years by Jim Nabors, brings an emotional tear from many of the fans.

Then come the most exciting words in sport today, "Gentlemen, start your engines," and thousands of horsepower are brought to life. The crews then scurry off the grid as the cars slowly start their roll behind the pace cars.

The rest is all up to the drivers as they attune their minds to the tremendous task ahead. Thirty-three drivers will vie for the huge Borg-Warner Trophy that has been sitting on pit row all month. There's also the thought of that huge pile of money (over $1.2 million for winner Rick Mears in 1991) and the international fame and fortune that comes from winning the world's most famous racing event.

There have been many memorable races during the

**Left:** *Camera crews and reporters surround the front row before the 1990 race.*

**Below left:** *Starting 33 cars in 11 rows of three each makes for an exciting scene, as the field roars into the first turn.*

**Bottom:** *Cars at racing speeds are supposed to stay above the white line at Indy. Cars that have dropped off the pace, or are entering or exiting the pits as John Andretti is doing here, are required to run below that line.*

**Overleaf:** *In 1989 action, Rocky Moran (33) leads Arie Luyendyk (9) through turn four early in the race. Moran would end up finishing 14th and Luyendyk 21st. Luyendyk would go on to record a surprising victory the following year.*

modern era where the winner wasn't decided until the final laps. Here is a year-by-year account, starting with the exciting competition in 1960.

The 1960 race has been called by many the best Indy 500 ever. There was a battle of major proportions up front, with Jim Rathmann winning over Indy legends Rodger Ward and Eddie Sachs.

The following year, a brash young driver by the name of A.J. Foyt won his first race, over Eddie Sachs, in a fabulous duel to the finish. Two late race stops, Foyt for fuel and Sachs for tires, kept the excitement boiling to the end before A.J. took the checkered. Fans and competitors mourned the death of popular driver Tony Bettenhausen, who was killed in a test accident.

Rodger Ward won his second 500 in 1962 with a close 11-second victory over teammate Len Sutton. Had not an

33
BANDIT

**Below:** *There are many viewing locations for the Indy 500. Two of the more unique spots are shown here – the elevated camera platform and the Goodyear Blimp.*

**Right:** *Not all of the action takes place on the track. A section inside the first turn, long known as the "Snake Pit," is where the young, and the young at heart, gather to picnic, drink beer, watch members of the opposite sex, and occasionally watch the race!*

**Below right:** *Al Unser, Sr. comes into the pits during the 1987 race. Pit stops are critically important to the outcome of the Indy 500. A mistake in the pits can negate any advantage the driver has worked lap after lap to achieve.*

error in the pits occurred, Foyt would have been in solid contention for this win.

The 1963 race was the start of the rear-engine revolution at Indy, and Jimmy Clark in one of those machines came very close to winning with a second place finish. But the front engine machines were still dominant, as Parnelli Jones took the victory.

Racing at Indy is dangerous, as the 1964 race proved in dramatic fashion. Drivers Dave McDonald and Eddie Sachs were killed in a front stretch inferno. The race was stopped for two hours. Parnelli Jones and A.J. Foyt duelled for 140 laps before Jones had a fuel tank problem in the pits and A.J. cruised to Indy victory number two.

The rear engine revolution arrived in force in 1965 when the first of the then-exotic machines, piloted by Jimmy Clark, totally monopolized Indy. It was the beginning of a new era for Indy. The following year, another Briton, rookie Graham Hill, won in a crash-filled affair.

It was quiet at Indy in 1967, but mostly because of the quiet whine of the turbine engine of the Andy Granatelli machines. With Parnelli Jones at the wheel, the jet car was far superior to the usual Indy machines and was on its way to victory when a tiny part failed with four laps to go, giving the race to Foyt. A.J. had to pick his way through a last lap crash on the main straightaway to seal his victory.

The turbine power was reduced for the 1968 race, but the turbine-powered cars still occupied two of the three front row positions. The jet engines failed late in the race, however, giving Bobby Unser the first of his three Indy checkereds.

Granatelli finally got his Indy day in the sun in 1969, but it was not with a turbine car. Hard-luck Indy driver Mario Andretti had crashed his primary car in practice, but when he got into his backup car, things started going his way. With all his challengers having mechanical problems, Mario took the checkered flag and also the famous kiss from owner Granatelli.

The Unser era continued in 1970 when a youthful Al Unser, Sr. took his first of four Indys with an easy win. A high finisher was stock car driver Donnie Allison, who finished fourth, and had just won the World 600 stock car race a day earlier.

Al Unser won again the following year for his second straight victory, but some of the excitement of this race took place on the pit lane as the green flag was thrown. The pace car could not get stopped after starting the race and crashed into a photographers' stand, injuring several people.

Mark Donohue would win his lone Indy in 1972 when Gary Bettenhausen, who had appeared to have the race won, broke down with only 18 laps to go. Al Unser, Sr. came close to making it three in a row with a second place finish.

What is possibly Indy's worst wreck ever, with many cars involved, occurred at the very start of the 1973 race. It was just the start of problems at the Brickyard that year, as consistent rains would finally stop and allow the race to be restarted two days later. Shortly after the race

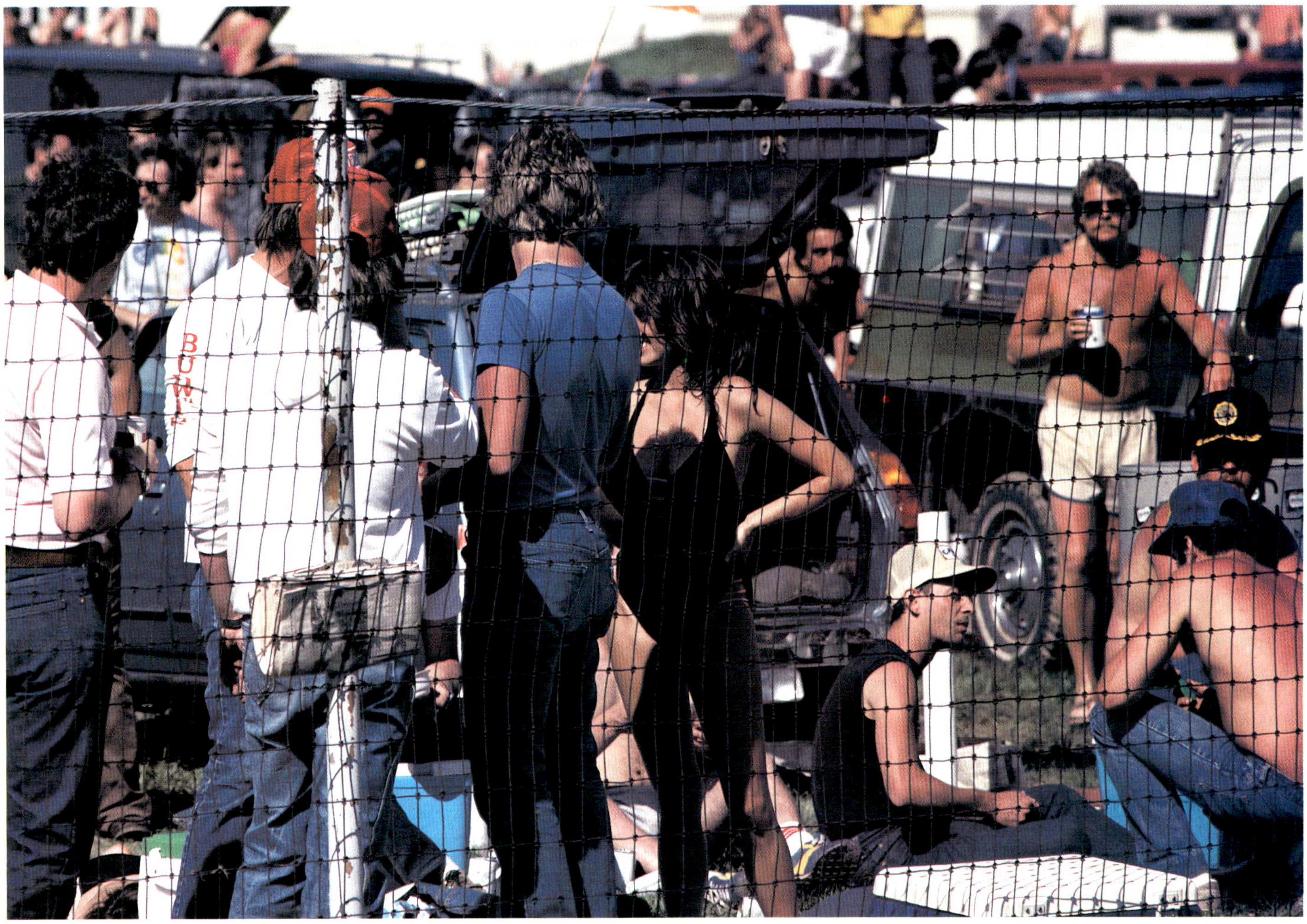

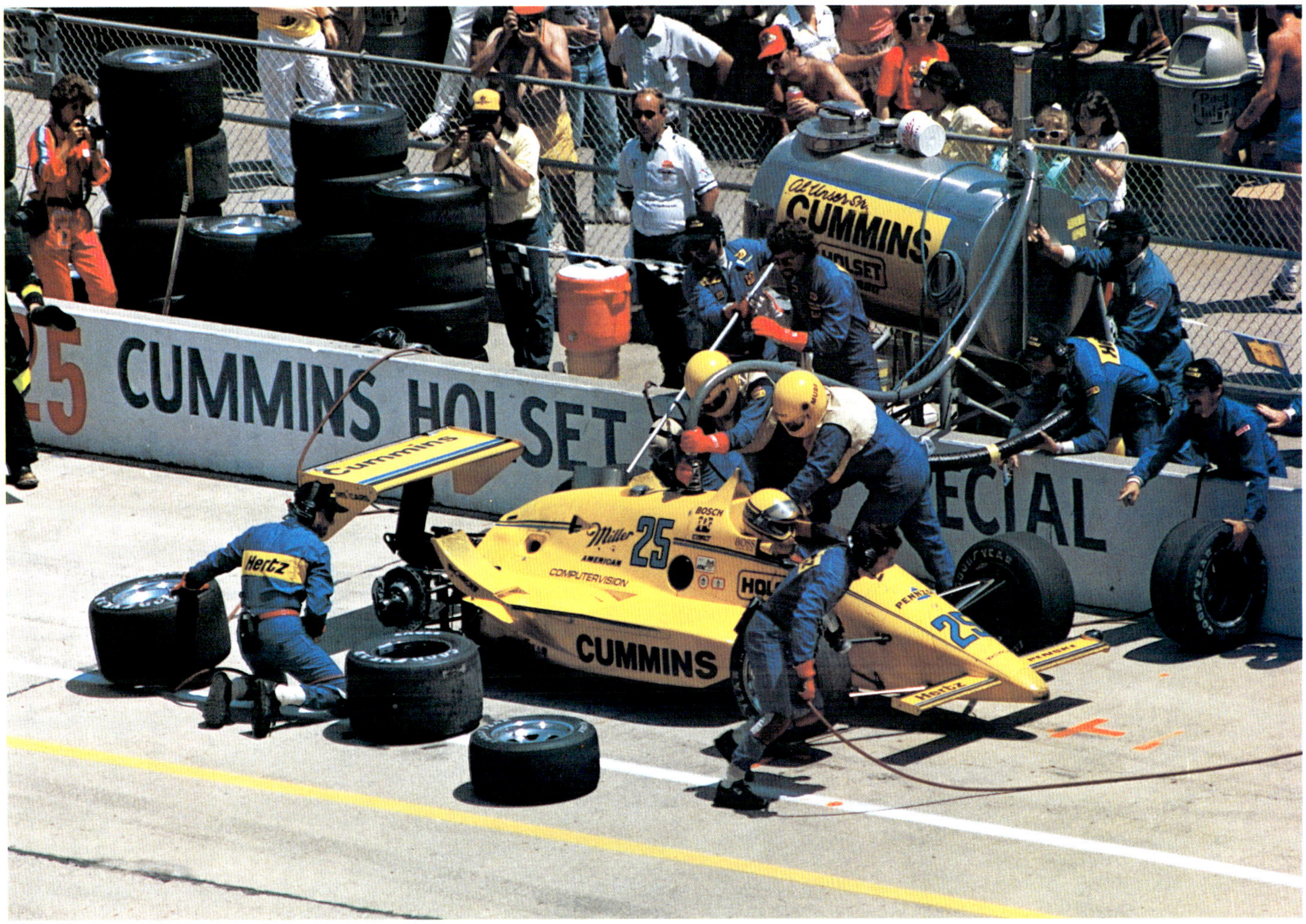
25 CUMMINS HOLSET
ECIAL
CUMMINS
HOLSET
Cummins
Miller
25
COMPUTERVISION
CUMMINS
Hertz

Budweiser 3
ARCIERO

**Left:** *With the speeds that are run at Indy, crashes are often spectacular. Here, Mario Andretti and Johnny Parsons collide with the first turn wall. This is definitely a white knuckle ride for both drivers, but fortunately, neither was hurt.*

**Overleaf:** *One of the favorite places on the track for a driver to pass the car ahead of him is in the short straightaways between the turns. Kevin Cogan (11) demonstrates this technique as he passes Dominic Dobson (86).*

start, driver Swede Savage was killed in a terrible crash. The victory finally went to Gordon Johncock.

One of the great Indy names, Johnny Rutherford, took the 1974 race. The pole had been won by Foyt, who challenged Rutherford before his turbocharger failed.

Bobby Unser's crew called the incoming rain and had their man ahead when the 1975 race was washed out with 26 laps to go. Rutherford and Foyt provided the close competition for this unfinished race.

Rain also played a large part in the next race when the race was barely half over, at 102 laps. Rutherford won again after starting on the pole. An interesting aspect of the race was that the first seven starters were all in the top seven at the finish.

Foyt would become the first four-time winner in 1977, a race which saw a woman, Janet Guthrie, qualify for the first time. As is so often the case at Indy, it appeared that a driver (Gordon Johncock in this instance) had the race won, but an engine failure late in the race put him on the sidelines.

The new Chaparral car design with Al Unser, Sr. aboard showed its superiority in 1978 as Al took his third win at Indy. But the high interest of this year was Tom Sneva's accomplishment of breaking the track record with a 202.156 clocking.

Battling on the sidelines in 1979 were the newly-formed CART organization and the long-time Indy sanctioning body, USAC. Both Bobby and Al Unser had a chance to win this race, and A.J. Foyt was in the race to the finish with a second place effort. Newcomer Rick Mears, driving for the Penske team, had replaced Ton Sneva and to the surprise of many, took the check ered flag.

Johnny Rutherford showed that the Chaparral was still a top design as he took the car to victory in 1980. The best effort of the race, though, was made by Tom Sneva, who started in last place after crashing his primary car and raced through the field for a second place finish.

There was a fiasco in scoring in the 1981 race when Bobby Unser was finally awarded the victory in October. After taking the checkered flag, Unser saw his victory taken away the following day when it was discovered that Bobby had passed cars under the yellow flag. It would be reversed again for a final time in October when Bobby got the victory back. It was a sad situation for Andretti, the "other" winner, who had driven the race of his life coming from 32nd position.

Kevin Cogan lost control of his front row machine in the 1982 race and knocked three cars from the race. But this race will always be remembered as one of the great Indy races. It boiled down to a duel between Gordon Johncock and Rick Mears. The crowd was on its feet as the final laps fell by and the two kept pushing each other. At the end, Johncock came across the line a scant .16

86
Havoline

86
BOSCH
11
TUNEUP MASTERS
BUICK
Valvoline
TUNEUP MASTERS
RCA
GOODYEAR
Delco Remy

**Below:** *To the victor goes the spoils: Al Unser, Sr. takes a happy ride with car owner Roger Penske following a tremendously unexpected victory in the 1987 race. Al came to Indy without a ride and left with his fourth win at the Brickyard.*

**Right:** *Emerson Fittipaldi happily obliges photographers on the day after his 1989 victory. The greatest feeling in the world must be to win the greatest race in the world – the Indy 500.*

seconds ahead of Mears to take the win. Many remember it as the most exciting Indy race ever.

A controversy hit the old track in 1983 when winner Tom Sneva insisted that he was being held up by Al Unser, Jr. as Sneva tried to get around Al Sr. Sneva finally did make the pass for his first victory.

In 1984, Rick Mears took his second Indy victory in a record-breaking run. Sneva had made a solid run for the lead, but the car broke down with 37 laps to go. The race marked the first race for rookie Michael Andretti, who finished fifth, and who would become an Indy Car contender in the late 1980s and early 1990s.

The famous "spin and win" race of 1985 had driver Danny Sullivan doing a complete 360-degree spin in

front of Mario Andretti. Fortunately, he hit nothing and went on to take the victory. Tough-luck Mario came close to that second Indy win, but had to settle for second in this one.

It came down to a three-car race in 1986 with Rick Mears falling off at the end for a third place finish. The final duel between Bobby Rahal and Kevin Cogan saw Rahal win the day. The winning pass was made with two laps to go, with the winning margin being a scant 1.4 seconds.

A car which had been used earlier in the year as a show car for the Penske team served as the winning mount for Al Unser, Sr. to take his fourth Indy win. Again, Andretti dominated early, but as has happened so many times to Mario, the car failed. Then, after coming from a lap down, Big Al came back and took the popular victory.

It was a matter of the old and the new for the 1988 race as Rick Mears in his familiar Pennzoil machine took his third win, and a driver that would be heard from again at Indy, Emerson Fittipaldi, finished second. It was car owner Roger Penske's fourth win in five years.

Make that five out of six for Penske, as Fittipaldi would show his superiority in 1989 and bring home the Indy win, to add to his Formula 1 World Championship. The Brazilian ace battled Al Unser, Jr. in a duel reminiscent of 1982, but this time only one of the cars would make it to the finish. During one of their close encounters, the cars touched and Little Al was spun out, leaving Fittipaldi to collect the honors.

With his mod dress and long hair, Arie Luyendyk was a breath of fresh air for the Indy set in 1990. He showed right away that he would be a contender with a front row qualifying effort, and then came back to prove his superiority with a 167th lap pass in the great race. At the end, he was ten seconds ahead of Bobby Rahal.

At Indy, you don't hold Roger Penske down very long, and the master returned to the winner's circle in 1991 with Rick Mears's fourth Brickyard victory. Needless to say, thoughts of a new record with a fifth win were sounded to Rick, but the future at Indy is always uncertain. The only certainty, in fact, is that the future holds many more thrill-packed races at one of the most exciting sporting events in the world – the Indy 500.

# RACE LOG

## '500' WINNERS

| Year | Winner | Year | Winner | Year | Winner |
|---|---|---|---|---|---|
| 1911 | Ray Harroun | 1938 | Floyd Roberts | 1967 | A.J. Foyt |
| 1912 | Joe Dawson | 1939 | Wilbur Shaw | 1968 | Bobby Unser |
| 1913 | Jules Goux | 1940 | Wilbur Shaw | 1969 | Mario Andretti |
| 1914 | Rene Thomas | 1941 | Floyd Davis & Mauri Rose | 1970 | Al Unser Sr. |
| 1915 | Ralph DePalma | 1946 | George Robson | 1971 | Al Unser Sr. |
| 1916 | Dario Resta | 1947 | Mauri Rose | 1972 | Mark Donohue |
| 1919 | Howdy Wilcox | 1948 | Mauri Rose | 1973 | Gordon Johncock |
| 1920 | Gaston Chevrolet | 1949 | Bill Holland | 1974 | Johnny Rutherford |
| 1921 | Tommy Milton | 1950 | Johnnie Parsons | 1975 | Bobby Unser |
| 1922 | Jimmy Murphy | 1951 | Lee Wallard | 1976 | Johnny Rutherford |
| 1923 | Tommy Milton | 1952 | Troy Ruttman | 1977 | A.J. Foyt |
| 1924 | Joe Boyer & L.L. Corum | 1953 | Bill Vukovich | 1978 | Al Unser Sr. |
| 1925 | Peter DePaolo | 1954 | Bill Vukovich | 1979 | Rick Mears |
| 1926 | Frank Lockhart | 1955 | Bob Sweikert | 1980 | Johnny Rutherford |
| 1927 | George Souders | 1956 | Pat Flaherty | 1981 | Bobby Unser |
| 1928 | Louis Meyer | 1957 | Sam Hanks | 1982 | Gordon Johncock |
| 1929 | Ray Keech | 1958 | Jimmy Bryan | 1983 | Tom Sneva |
| 1930 | Billy Arnold | 1959 | Rodger Ward | 1984 | Rick Mears |
| 1931 | Louis Schneider | 1960 | Jim Rathmann | 1985 | Danny Sullivan |
| 1932 | Fred Frame | 1961 | A.J. Foyt | 1986 | Bobby Rahal |
| 1933 | Louis Meyer | 1962 | Rodger Ward | 1987 | Al Unser Sr. |
| 1934 | Bill Cummings | 1963 | Parnelli Jones | 1988 | Rick Mears |
| 1935 | Kelly Petillo | 1964 | A.J. Foyt | 1989 | Emerson Fittipaldi |
| 1936 | Louis Meyer | 1965 | Jimmy Clark | 1990 | Arie Luyendyk |
| 1937 | Wilbur Shaw | 1966 | Graham Hill | 1991 | Rick Mears |

## TEN MILE PER HOUR INCREASES IN ONE-LAP QUALIFYING SPEEDS

| YEAR | SPEED | DRIVER | BARRIER |
|---|---|---|---|
| 1914 | 94.54 miles per hour | Rene Thomas | 90 mph |
| 1919 | 104.78 | Rene Thomas | 100 |
| 1925 | 110.728 | Earl Cooper | 110 |
| 1927 | 120.546 | Peter DePaolo | 120 |
| 1937 | 130.492 | Jimmy Snyder | 130 |
| 1954 | 141.287 | Jack McGrath | 140 |
| 1962 | 150.729 | Parnelli Jones | 150 |
| 1965 | 160.973 | Jim Clark | 160 |
| 1968 | 171.887 | Graham Hill | 170 |
| 1972 | 185.797 | Bill Vukovich | 180 |
| 1972 | 196.678 | Bobby Unser | 190 |
| 1977 | 200.535 | Tom Sneva | 200 |
| 1984 | 210.689 | Tom Sneva | 210 |
| 1988 | 220.453 | Rick Mears | 220 |

## POLE POSITION WINNERS

| YEAR | DRIVER | ENGINE |
|---|---|---|
| 1911 | Louis Strang | Case |
| 1912 | Gil Anderson | Stutz |
| 1913 | Caleb Bragg | Mercer |
| 1914 | Jean Chassagne | Sunbeam |
| 1915 | Howdy Wilcox | Stutz |
| 1916 | Johnny Aitken | Peugeot |
| 1919 | Rene Thomas | Ballot |
| 1920 | Ralph DePalma | Ballot |
| 1921 | Ralph DePalma | Ballot |
| 1922 | *Jimmy Murphy | Miller |
| 1923 | *Tommy Milton | Miller |
| 1924 | Jimmy Murphy | Miller |
| 1925 | Leon Duray | Miller |
| 1926 | Earl Cooper | |
| 1927 | Frank Lockhart | Miller |
| 1928 | Leon Duray | Miller |
| 1929 | Cliff Woodbury | Miller |
| 1930 | *Billy Arnold | Miller |
| 1931 | Russ Snowberger | Studebaker |
| 1932 | Lou Moore | Miller |
| 1933 | Bill Cummings | Miller |
| 1934 | Kelly Petillo | Offy |
| 1935 | Rex Mays | Offy |
| 1936 | Rex Mays | Offy |
| 1937 | Bill Cummings | Offy |
| 1938 | *Floyd Roberts | Offy |
| 1939 | Jimmy Snyder | Sparks |
| 1940 | Rex Mays | Sparks |
| 1941 | Mauri Rose | Maserati |
| 1946 | Cliff Bergere | Offy |
| 1947 | Ted Horn | Maserati |
| 1948 | Rex Mays | Bowes |
| 1949 | Duke Nalon | Novi |
| 1950 | Walt Faulkner | Offy |
| 1951 | Duke Nalon | Novi |
| 1952 | Fred Agabashian | Cummins |
| 1953 | *Bill Vukovich | Offy |
| 1954 | Jack McGrath | Offy |
| 1955 | Jerry Hoyt | Offy |
| 1956 | *Pat Flaherty | Offy |
| 1957 | Pat O'Connor | Offy |
| 1958 | Jim Rathmann | Offy |
| 1959 | Johnny Thomson | Offy |
| 1960 | Eddie Sachs | Offy |
| 1961 | Eddie Sachs | Offy |
| 1962 | Parnelli Jones | Offy |
| 1963 | *Parnelli Jones | Offy |
| 1964 | Jim Clark | Ford |
| 1965 | A.J. Foyt, Jr. | Ford |
| 1966 | Mario Andretti | Ford |
| 1967 | Mario Andretti | Ford |
| 1968 | Joe Leonard | Turbine |
| 1969 | A.J. Foyt, Jr. | Ford |
| 1970 | *Al Unser, Sr. | Ford |
| 1971 | Peter Revson | Offy |
| 1972 | Bobby Unser | Offy |
| 1973 | Johnny Rutherford | Offy |
| 1974 | A.J. Foyt, Jr. | Foyt |
| 1975 | A.J. Foyt, Jr. | Foyt |
| 1976 | *Johnny Rutherford | Offy |
| 1977 | Tom Sneva | Cosworth |
| 1978 | Tom Sneva | Cosworth |
| 1979 | *Rick Mears | Cosworth |
| 1980 | *Johnny Rutherford | Cosworth |
| 1981 | *Bobby Unser | Cosworth |
| 1982 | Rick Mears | Cosworth |
| 1983 | Teo Fabi | Cosworth |
| 1984 | Tom Sneva | Cosworth |
| 1985 | Pancho Carter | Buick |
| 1986 | Rick Mears | Cosworth |
| 1987 | Mario Andretti | Chevy |
| 1988 | *Rick Mears | Chevy |
| 1989 | Rick Mears | Chevy |
| 1990 | Emerson Fittipaldi | Chevy |
| 1991 | *Rick Mears | Chevy |

* Indicates Race Winner

# INDEX

Numbers in *italics* indicate illustrations